Journal of the Geological Society, London, Vol. **148**, 1991, pp. 813–824, 5 figs, Printed in Northern Ireland

A guide to stratigraphical procedure

A. WHITTAKER[1], J. C. W. COPE[2], J. W. COWIE[3], W. GIBBONS[2], E. A. HAILWOOD[4], M. R. HOUSE[5], D. G. JENKINS[6], P. F. RAWSON[7], A. W. A. RUSHTON[1], D. G. SMITH[8], A. T. THOMAS[9] & W. A. WIMBLEDON[10].

[1] *British Geological Survey, Keyworth, Nottingham NG12 5GG, UK*

[2] *Department of Geology, University of Wales, PO Box 914, Cardiff CF1 3YE, UK*

[3] *Department of Geology, The University, Bristol BS8 1RJ, UK*

[4] *Department of Oceanography, The University, Southampton SO9 5NH, UK*

[5] *Department of Geology, The University, Southampton SO9 5NH, UK*

[6] *The Old Chapel, Rhossili, Swansea SA3 1PH, UK*

[7] *Department of Geological Sciences, University College London, Gower Street, London WC1E 6BT, UK*

[8] *BP International Ltd, Sunbury Research Centre, Chertsey Road, Sunbury-on-Thames, Middlesex TW16 7LN, UK*

[9] *Department of Geological Sciences, University of Birmingham, PO Box 363, Birmingham B15 2TT, UK*

[10] *Nature Conservancy Council, Northminster House, Peterborough PE1 1UA, UK*

Abstract: The continuing development of stratigraphical concepts and the application of new methods has led the Stratigraphy Committee of the Geological Society of London to present a new guide to stratigraphical procedure. Some of the principal types of stratigraphy are outlined and guidelines and recommendations are given on procedure, practice and terminological usage.

Stratigraphy is the study and description of rock successions. It leads to the interpretation and fuller understanding of rock successions as sequences of events in the geological history of the Earth. This account, presented by the Geological Society's Stratigraphy Committee, is basically concerned with guidance on the establishment of unambiguous and consistent procedure, terminology and nomenclature in stratigraphical description, analysis and communication. These matters are important because stratigraphy provides methods of analysis and interpretation central to many fields of geological investigation, for example in basin evolution and dynamics. Well-founded stratigraphical procedures and principles promote systematic and rigorous study of the composition, geometry, sequence, history and genesis of rocks. They provide the framework within which temporal and spatial relationships among rock bodies are ordered systematically.

Although these factors provide the justification for stratigraphical classification, it is important that classification does not become an end in itself. The contributors to this guide recognize the potential danger of inhibiting new approaches to, and concepts in, stratigraphy should codes and rules be applied too rigidly. This, we think, could be detrimental to the longer term progress of stratigraphy. Accordingly, this account, like its predecessor (Holland *et al.* 1978), does not try to codify stratigraphical nomenclature, although it does attempt to be helpful to those who need to use stratigraphical classification and terminology.

The fundamental procedure in stratigraphy is to observe, describe and correlate rock successions, correlation being the demonstration of correspondence between geological units in some defined property and in relative stratigraphical position. When investigating an area, the first procedure is to describe the local stratigraphical succession objectively in terms of the mappable or traceable lithostratigraphical units present. Concurrently or subsequently the local succession should be correlated, as far as is possible, with the international standard stratigraphical scale or with regional stratigraphical divisions. With most Phanerozoic rocks this correlation is most commonly achieved using biostratigraphical methods.

Three principal approaches to stratigraphy (lithostratigraphy, biostratigraphy, chronostratigraphy) are widely recognized. Other methods (e.g. magnetostratigraphy, geophysical log stratigraphy, seismic stratigraphy etc.) augment stratigraphical practice. Holostratigraphy is the combined use of all possible methods. Some of these methods and their terminology are outlined below.

A separate procedure is the actual dating of rocks, or of tectonic and metamorphic effects expressed in them, by the methods of geochronometry. In the earlier Precambrian in particular, geochronometry has a major role in correlation of radiometrically dated successions with the international stratigraphical standard.

Lithostratigraphy

The description, definition and naming of rock units is termed lithostratigraphy (rock stratigraphy). It is fundamental to all other branches of stratigraphy because accuracy in, for example, biostratigraphy and chronostratigraphy relies on the correct recognition of the spatial relationships (both vertical and lateral) of rock units.

Individual units are described and defined using their gross lithological characteristics and by their interrelationships with adjacent units. Gross lithological characteristics embrace petrography, mineralogy, geochemistry, general fossil content etc. Units are named according to their perceived rank in a formal lithological hierarchy (see Fig. 1), though that rank can be changed subsequently (e.g. the Brent sands of the northern North Sea were initially described as the Brent Formation but have since been reclassified as the Brent Group).

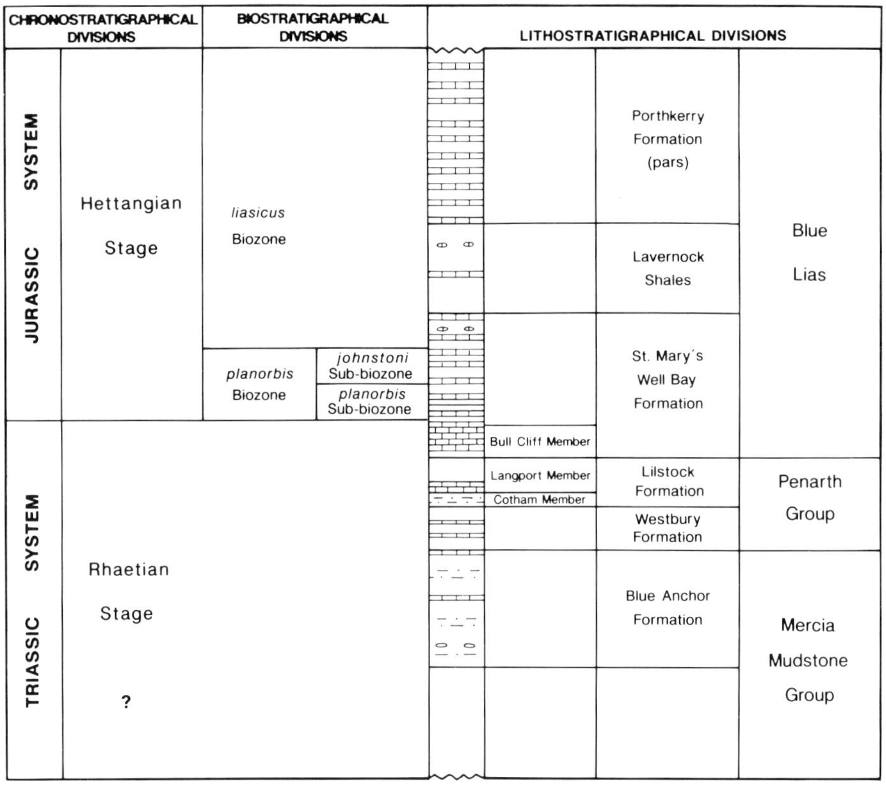

Fig. 1. Stratigraphical divisions near the Triassic–Jurassic boundary, Penarth area, South Wales.

Establishment of lithostratigraphical units

When proposing a new lithostratigraphical unit (or revising an old-established one), it is necessary to be as precise as possible. A newly established or revised lithostratigraphical unit requires the following information to be published to give formal validity to the named unit. (1) The exact location of its type section; (2) the unit's distinguishing lithological characteristics including those of its lateral variations; (3) its relationship to other units in the overall succession; (4) its thickness in the type and other sections, and in its lateral variations; (5) its relationship to pre-existing classifications. The defined unit should have or should once have had physical continuity. Description should include especially how the base of the unit is recognized and also as appropriate, the mineralogical and palaeontological content of the unit.

The lithostratigraphical hierarchy

The basic hierarchy consists of:

- supergroup
- group
- formation
- member
- bed

The **formation** is the basic unit of the lithostratigraphical hierarchy. It is commonly defined as the smallest mappable or traceable unit, though that depends to some degree on the scale of mapping. It generally varies in maximum thickness from a few metres to several hundreds of metres.

A formation has lithological characteristics that distinguish it from adjacent formations. Its boundaries may be sharp, or fixed arbitrarily in a gradational sequence. The lithology may be broadly similar throughout the formation, but there may be some quite distinctive sub-units that are best regarded as members. For example, the widely distributed Kimmeridge Clay Formation (Jurassic) is usually argillaceous throughout, but in South Humberside it contains a local sandstone member, the Elsham Member, and in south Dorset an ironstone, the Abbotsbury Member. Likewise a gross lithological characteristic is not necessarily a uniform one: the Cleveland Ironstone Formation (Jurassic) of North Yorkshire embraces several coarsening-upwards cycles (shales to siltstones to ironstones) which clearly distinguish it from the sandy Staithes Formation below and the argillaceous Whitby Mudstones Formation above.

A formation should be laterally continuous or have once been so. Individual formations are commonly limited to a single sedimentary basin and the extension of the name to adjacent basins may be debatable. A good example is provided again by the Kimmeridge Clay Formation, named originally from Dorset, in the Wessex Basin. The name was then adopted for similar clays in the Yorkshire Basin, with justification because the outcrops are virtually continuous. The name has since been applied indiscriminately to offshore basins as far away as Norway with confusing results (see Doré *et al.* 1985).

A formation may stand alone in the lithostratigraphical hierarchy or it may be classified together with adjacent formations to form a **group**, but the definition of a group can be contentious. Some stratigraphers insist that it must have a dominant lithological characteristic e.g. the Lias Group is predominantly argillaceous though containing sandstone- and ironstone-dominated formations. Others place more emphasis on the genesis of the group, defining it more by its boundaries than by its internal characteristics. Ideally both approaches should be taken into account, but there is no doubt that with the burgeoning application of geophysical techniques to stratigraphical practice the occurrence of major bounding unconformities will play an increasingly important role in definitions. In any case, such unconformities often mark substantial facies changes, though a group located between two major unconformities may embrace numerous lesser stratigraphical breaks and several lithologically distinctive formations.

A **member** is a subdivision of a formation. Several members may be contained within a formation and characterized by distinctive lithologies. Members commonly form a wedge-shaped entity within the formation, such as a marginal sand body prograding into a predominantly argillaceous basin.

The smallest formal unit is a **bed**. It is rarely worth giving a formal name to every individual bed in a rapidly varying sequence, but in lithologically monotonous sequences such as the Chalk Group, formal bed names for distinctive fossil, marl and flint bands may have considerable reference value (e.g. Wood & Smith 1978).

Naming of lithostratigraphical units

Each formally defined new lithostratigraphical unit should have a name including a proper name, chosen where possible from distinctive geographical features near a characteristic section, ideally a type section. The names should be chosen for uniqueness and for convenience in usage. The same geographical name should not be used for more than one unit. Names of lithostratigraphical units and their spelling, as defined originally, should not normally be changed even if the eponymous name changes. The inclusion of a single lithological term in a new name is recommended but is not mandatory where a single lithological term would be inappropriate. All the initial letters of formally defined lithostratigraphical terms are capitalized, as in Sherwood Sandstone Group. For new designations, additional terms such as lower, middle and upper should be used informally only and thus without initial capital in these words.

Stability in stratigraphical nomenclature

Long-established stratigraphical nomenclature may not fit readily with the tenets expressed above, and if used lithostratigraphically there may be confusion with modern chronostratigraphical terms. Clearly if major new concepts are incorporated in the stratigraphical revision of a previously described area, new names will be called for; but on the whole continuity and understanding are best served if well-established concepts retain their familiar names, whilst being accorded modern definitions, type sections and so on. Even when major definitive revision is undertaken it may be advantageous to retain a reference to the old name, e.g. when the (lithostratigraphical) Wenlock Limestone was revised, a clash with the (chronostratigraphical) Wenlock Series was avoided by renaming the former the Much Wenlock Limestone Formation. Changes of name to avoid homonyms (e.g. there are several examples of 'White Limestone') are seen as mandatory, whereas others may be desirable if they are lithological in character ('Chert Beds'). Other irregularly-formed names should be conserved so long as confusion does not arise (e.g. Cornbrash, Clypeus Grit, MacGregor Marine Band).

Biostratigraphy

Biostratigraphy is the use of fossils for stratigraphical correlation and the organization of strata into biostratigraphical units on the basis of their fossil content (Hancock 1977); such a unit is termed a **biozone**.

Because successions of fossils have been found to be broadly similar over wide areas at certain times, it is clear that the faunal and floral successions thus displayed are chronologically arranged. So, for example, successions of graptolite faunas in the Lower Palaeozoic or of ammonoids in the Upper Palaeozoic and Mesozoic, allow one to effect a fine correlation which clearly has temporal significance. Biostratigraphy thus provides a means by which a chronostratigraphical scale can be constructed and applied to lithological successions.

Biozones are often named after one or more fossil taxa (usually species) which constitute the **index fossils**. The name of the index fossil should always appear in italics. Following binomial practice, the initial letter of the generic name should be capitalized, whilst the initial letter of the specific name should be a lower case letter. It is not always necessary to refer to the whole binomial name when referring to a biozone (particularly for a second or subsequent reference to the same biozone) but the same rules apply. Thus the *Glyptograptus teretiusculus* Biozone may be referred to as the *teretiusculus* Biozone. Note, too, the capitalization of the word 'Biozone'. Fossil names used as biozonal indices are subject to the International Rules of Zoological (Ride *et al.* 1985) or Botanical (Greuter *et al.* 1988) Nomenclature and should thus be validly published taxa. The biozonal name is subject to change should the name of the index species be changed.

Nektonic or planktonic organisms are more likely to provide satisfactory biozonal indices, since they are largely independent of facies. Thus groups such as graptoloids, ammonoids or planktonic foraminifera have proved to be reliable zonal indices, whereas benthic forms such as brachiopods, corals or some trilobites, being to a greater extent facies controlled, have proved less reliable for correlation.

Biozones may be of any thickness (from a single thin bed to units thousands of metres thick) and their duration may also vary depending on the nature and stratigraphical longevity of the index fossil(s). Thus, for the Jurassic, ammonite biozones probably averaged around 600 000 years in duration (Cope *et al.* 1980*a*, p. 17) whilst similar orders of duration may also apply to Upper Cambrian olenid trilobite biozones. On the other hand, as pointed out by Holland *et al.* (1978, p. 13) some Lower Cambrian trilobite biozones may have had a duration as long as ten million years.

A **sub-biozone** is a subdivision of a biozone. An even smaller unit, the **biohorizon**, may be recognized locally or regionally, within or at the boundaries of biozones; such units can give even greater precision in correlation. As an example of a biohorizon, beds of very limited vertical extent containing species of the ammonite *Nannocardioceras* have been recognized in Kimmeridgian (Upper Jurassic) strata from Dorset to the Weald and to Yorkshire (Cope *et al.* 1980*b*, Kimmeridgian Chart). These *Nannocardioceras*-yielding beds occur within the *Aulacostephanus eudoxus* Biozone and form a very restricted biohorizon providing a very precise correlation at this level.

Biozones may be of different types (see Fig. 2) and the

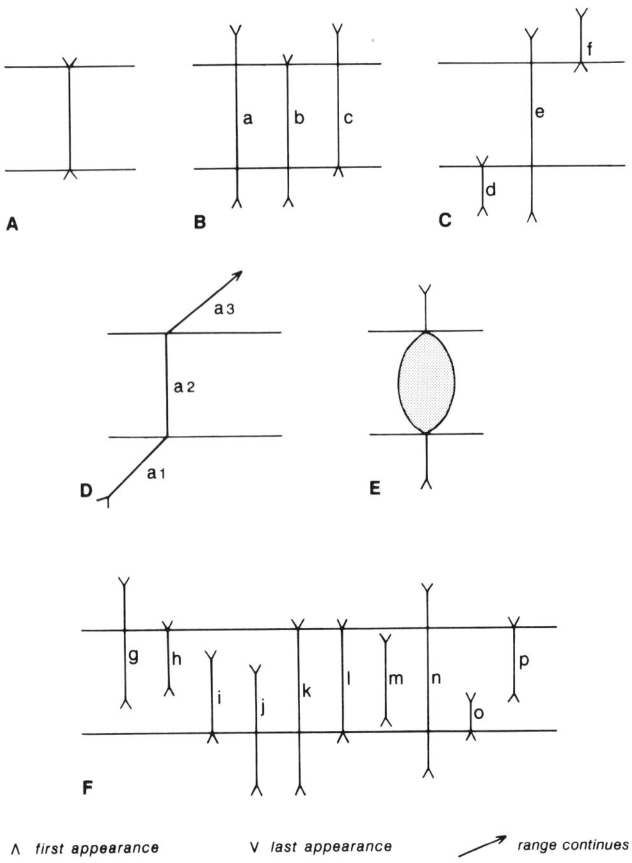

Fig. 2. Different types of biozones. (**A**) Total Range (Local Range) Biozone; (**B**) Concurrent Range Biozone defined by overlap of taxa a, b, c; (**C**) Partial Range Biozone; (**D**) Consecutive Range Biozone (species a1, a2, and a3 are phylogenetically linked); (**E**) Acme Biozone; (**F**) Assemblage Biozone.

principal ones are described below. Many biozones are based on the stratigraphical range of the index taxon (usually species). Such a major category of biozone is the **Range Biozone**; this may be subdivided into several types. A **Total Range Biozone** encompasses strata containing the total stratigraphical and geographical range of the index taxon; such biozones are difficult to justify over more than local areas and thus become **Local Range Biozones**. Both these types of biozone are defined by the appearance and disappearance of the nominal taxon; in practice, disappearances can be difficult to establish and it is good stratigraphical practice to define a biozone only by its base, thus allowing the top of the zone to be fixed by the base of the succeeding biozone. In much borehole work, however, the upper limit of a biozone is defined by the first downhole occurrence of the nominal taxon. A **Concurrent Range Biozone** is established by reference to the overlapping stratigraphical ranges of several taxa. A **Partial Range Biozone** is established within the stratigraphical range of a taxon, part of the range lying above that of another taxon and part below that of a further taxon (Fig. 2). In the case of an established phylogenetic lineage, it is possible to isolate the range of an intermediate species in such a lineage as a **Consecutive Range Biozone**.

A further type of biozone is the **Acme Biozone**; this relies on the fact that a particular taxon may, for a particular period of time, be extraordinarily abundant. Biozones founded in this way are likely to be particularly subjective and of local application only. There are real problems in an objective definition of the acme of a species, and there remains the probability that local environmental factors may have been the cause of a local acme, which may be unrelated to an acme of the same species elsewhere.

Also used are **Assemblage Biozones**, which are biozones characterized by an assemblage of taxa which, taken together, constitute an assemblage which differs in its faunal or floral character from contiguous assemblages. An assemblage biozone cannot be strictly defined; problems in definition become apparent on examination of Fig. 2F, which is an example of the type of diagram sometimes used in illustration of an assemblage biozone. Thus the taxa shown in Fig. 2F can be used in different ways to define the biozone: for example the biozone equals the total range of species l and thus could be defined as the total range biozone of that species. The discovery of species j on its own would not allow recognition of the biozone, but a combination of species j with species g would do so. Similarly, co-occurrence of species g and i would allow recognition of this biozone. Note that in this example we can specify what allows *recognition* of the biozone, but we cannot so easily arrive at a *definition* of the biozone (see discussion of this issue by Temple (1988) and Holland & Bassett (1988)). Despite these problems, assemblage biozones are useful when other methods of biozonation are not available. For example, the coral/brachiopod biozones of the Carboniferous Limestone of southern Britain, introduced by Vaughan (1905), are assemblage biozones which have considerable local value and provide the best macrofossil correlations yet achieved in that area. The Vaughan scheme of biozones could not be applied to northern England where the ranges of several species were found to be markedly different. A useful example of the practical definition and recognition of various kinds of biozones is set out in Woollam & Riding (1983).

Chronostratigraphy

Chronostratigraphy is the definition of internationally agreed boundaries for systems, series and stages. Boundaries of chronostratigraphical divisions should be defined in a designated boundary stratotype on the basis of observable biological or non-biological features of the rocks. Only at a defined **Global Stratotype Section and Point** (GSSP) (e.g. at the base of the Devonian) can a point in rock be known with certainty to indicate specified time. Elsewhere, correlation with the chronostratigraphical scale is achieved with the maximum refinement possible, using all available criteria.

The necessary definition of the base of a standard chronostratigraphical division is achieved at a marker point (the 'golden spike') in the boundary stratotype section (Holland 1986). The top of the division is defined automatically by definition of the base of the division above. The International Union of Geological Sciences, through its Commission on Stratigraphy and various Subcommissions, will arrange for the investigation, discussion, selection, definition, and description of such global boundary stratotype sections and points and for their international ratification and promulgation. Conservation and reasonable (though perhaps controlled) accessibility of boundary stratotype sections are essential.

New chronostratigraphical divisions require new names, preferably geographical in origin and related to the type area within which the basal boundary stratotype is chosen. Acceptable existing names of chronostratigraphical divisions may continue to be employed, though the newly defined basal boundary stratotype and the standard area in which it lies may be different from the original eponymous type area. Thus the Devonian System is named after Devonshire (south western England), but the base of the Devonian is now defined at a boundary stratotype in Czechoslovakia and the top by the basal Carboniferous boundary stratotype in southern France. Both proper name and hierarchical term of a chronostratigraphical division when used together require an initial capital letter, e.g. Devonian System.

A chronostratigraphical hierarchy as shown below is now widely recognized. The principal terms for corresponding divisions of geological time (geochronological divisions) are placed alongside the chronostratigraphical terms; e.g. rocks of the Permian System were deposited during the Permian Period.

Chronostratigraphical divisions	*Corresponding geological time divisions*
eonothem	eon
erathem	era
system	period
series	epoch
stage	age

Subdivisions of these principal chronostratigraphical divisions are in use. At the erathem level, for example, the Palaeozoic Erathem is subdivided into lower and upper parts with the boundary at the base Devonian GSSP. At the system level, the Carboniferous Subcommission of the International Commission on Stratigraphy has recently divided the Carboniferous System into two (as yet unnamed) subsystems with a mutual boundary at the middle of the Chokierian Stage. These two subsystems are *not* equivalent to Dinantian–Silesian (Cowie & Bassett 1989).

Examples of internationally agreed chronostratigraphical divisions are shown in Fig. 3 (see also Fig. 5). Both the Silurian System and the Devonian System are referred to the Palaeozoic Erathem which, in turn, is referred to the Phanerozoic Eonothem (Cowie & Bassett 1989).

The definition of boundary stratotypes for systems is of prime importance. Boundaries between systems must by definition coincide with the appropriate boundaries of divisions of lower rank: the series and stages. Boundaries of series by definition coincide with the appropriate boundaries of stages; boundaries of stages by definition coincide with the GSSP. It should be emphasized that the series is not a lithostratigraphical division, though it has often been used as such in the past, e.g. 'Borrowdale Volcanic Series' has been replaced by 'Borrowdale Volcanic Group'. The stage is widely regarded as especially useful in regional and global synthesis. For distinctiveness the ending -ian should be employed for names of stages, e.g. Campanian Stage, but not for those of series, e.g. Tremadoc Series.

SYSTEM	SERIES	STAGE
★ DEVONIAN	UPPER	Famennian
		Frasnian
	★ MIDDLE	Givetian
		Eifelian
	★ LOWER	Emsian
		Pragian
		★ Lochkovian
★ SILURIAN	PŘIDOLÍ	
	★ LUDLOW	Ludfordian
		★ Gorstian
	★ WENLOCK	Homerian
		★ Sheinwoodian
	★ LLANDOVERY	Telychian
		★ Aeronian
		★ Rhuddanian
★		

★ = *Global Stratotype Section and Point (GSSP)*

Fig. 3. Examples of some internationally agreed chronostratigraphical divisions.

Geochronometry

Geochronometry is the application of techniques which attempt to measure the age of rocks in years. The geochronometrical scale is a periodic scale using the year as the basic unit. Apparent ages obtained in geochronometry are referred to as radiometric or isotopic dates. For older rocks, multiple annual units are recommended to be written in millions of years (Ma), or in the case of the oldest rocks, in thousands of millions of years (Ga); Holocene and Pleistocene dates are normally quoted in years before 1950 (years BP (before present)).

Rank terms of geological time (eon, era, period, epoch and age) may be used for geochronometrical units when such terms are formalized. For example, the Archaean Eon and the Proterozoic Eon could be formal geochronometrical units distinguished on the basis of an arbitrarily chosen boundary at 2.5 Ga (2500 Ma).

Magnetostratigraphy

Magnetostratigraphy utilizes stratigraphical variations in the magnetic properties of rocks as a basis for geological correlation (Hailwood 1989). The two most useful properties are (i) the magnetic susceptibility, which is controlled by the composition and concentration of magnetic minerals in the rock and (ii) the direction of remanent magnetism, which is controlled by the direction (particularly the polarity) of the geomagnetic field at the time of formation or magnetization of the rock. The former can provide a valuable supplement to other lithostratigraphical data, while the latter gives the opportunity for correlating sedimentary or other rock sequences with the radiometrically-dated magnetic polarity time scale. This time scale is derived primarily from combined palaeomagnetic/radiometric dating analyses of volcanic lava sequences, supplemented by ocean-floor spreading magnetic anomaly data. Such correlations provide an important basis for the chronometric calibration of biostratigraphical zonations defined in sedimentary sequences, without the specific need for radiometric age determinations on the sediments themselves.

The basic magnetostratigraphical rock unit is the **magnetozone** which generally is identified as a rock interval characterized by a specific (either normal or reverse) magnetic polarity. However, it may be defined also in terms of its magnetic susceptibility or some other distinctive magnetic property. The interval of time corresponding to a particular magnetic polarity zone is the **magnetic chron**, which has a typical duration of about 10^5 to 10^6 years. Shorter intervals of opposing polarity within a chron are called **subchrons** and longer intervals of dominantly normal, reverse or mixed polarity are called **superchrons**.

The four most recent magnetic chrons have been named after pioneering workers in geomagnetism (which in younger to older order are: the Brunhes normal, Matuyama reverse, Gauss normal and Gilbert reverse polarity chrons), but earlier chrons conventionally are labelled from the corresponding ocean-floor spreading magnetic anomaly number.

Since geomagnetic polarity reversals are believed to be globally synchronous events and to be effectively 'instantaneous' on a geological time scale, polarity transitions should represent true time planes within rock sequences and provide an important basis for global geological correlations.

Borehole data and geophysical log stratigraphy

The procedures for establishing formal stratigraphical units utilizing data observed and described at the Earth's surface also apply to data acquired in the subsurface, both onshore and offshore. Complete lithological and palaeontological descriptions or logs of the cuttings or core samples are required in written or graphic form, or both. Accurate geographical location information is required. Boundaries of units should be indicated with their depths from an established reference datum. Naming of stratigraphical units can be taken from nearby geographical features. Geophysical log data are used in combination with lithological information from samples to produce a composite log from which type well sections can be selected and used for stratigraphical reference and correlation (see Rhys 1974, 1975; Deegan & Scull 1977).

Although subsurface stratigraphical description and nomenclature use lithostratigraphy and biostratigraphy, the scarcity of sample material from borehole sources and difficulties of precise sample depth location commonly necessitate that geophysical log data assume a place of great importance. Additionally, and importantly, the data can be used in a 'stand-alone' way even if no sample material is available. The geophysical log data comprise a continuous, 'in situ' record of the physical (and chemical) properties of the rock sequence drilled through. The log data provide lithological information, accurate location of bed boundaries and sequence information, and thus permit correlation of strata (Whittaker *et al.* 1985).

Even when compared with visually described cored sections, the geophysical log data from the same interval reveal features that cannot be detected by eye. Because of this the log data greatly augment stratigraphical studies and provide additional information for correlation. The log data enable more objective definition of sequences; they allow repeatable, objective description, and they are amenable to mathematical treatment, signal analysis and automatic correlation.

In addition, borehole velocity and density logs are used in the accurate geological calibration and interpretation of seismic reflection profiles. They enable depth calculation and the geological identification of individual seismic reflection events.

Seismic stratigraphy

Seismic stratigraphy is the interpretation of seismic reflection profiles in terms of the chronostratigraphical history of a sedimentary basin. It relies on the observation that reflections generally parallel stratal surfaces, i.e. they follow bedding surfaces that are assumed, at the vertical resolution of the seismic section, to represent a fixed interval of geological time in the depositional history of the rock column. Seismic reflection profiles are particularly useful for interpreting the large-scale accumulation history of a basin, but suffer with regard to information on lithology, which has to be arrived at indirectly.

Seismic reflections are commonly observed to exhibit both continuity and conformity within 'packages'. Unconformable relationships are found at reflector terminations; they show a variety of styles (comparable with those recognized at outcrop), including truncation, onlap and downlap. The foundation of the seismic stratigraphical method is the recognition of these unconformity-bounded packages on the profile as representing discrete depositional episodes in the history of the basin. It is found to be a reasonable working hypothesis that deposition of packages above and below these unconformities was entirely non-overlapping in geological time; i.e. the unconformities have direct chronostratigraphical significance. On this basis, the geological history represented by a seismic profile may be built up into a chronostratigraphical chart as illustrated in Fig. 4.

Terminology for the elements of seismic stratigraphical analysis is not mandatory, but that recommended below for sequence stratigraphy is appropriate and is in fairly general use (see Wilgus *et al.* 1988). Terms that illustrate the methodology of seismic stratigraphy, though they are too new to be recommended on the basis of usage, may be found in Vail (1987) and Van Wagoner *et al.* (1987). The hypothesis that depositional sequences correlate chronostratigraphically from basin to basin and worldwide is one that is entirely separate from the seismic stratigraphical method, even though the two first came to prominence in the same publication (Payton 1977). The Global Sedimentary Geology Programme of the International Union of Geological Sciences (Anon 1988) is studying these problems and others through a holostratigraphical approach.

Sequence stratigraphy

The concept of the depositional sequence was introduced by Sloss (1963) to describe the history of cratonic North America in terms of a succession of stratigraphical packages that are internally very varied, but unified by being

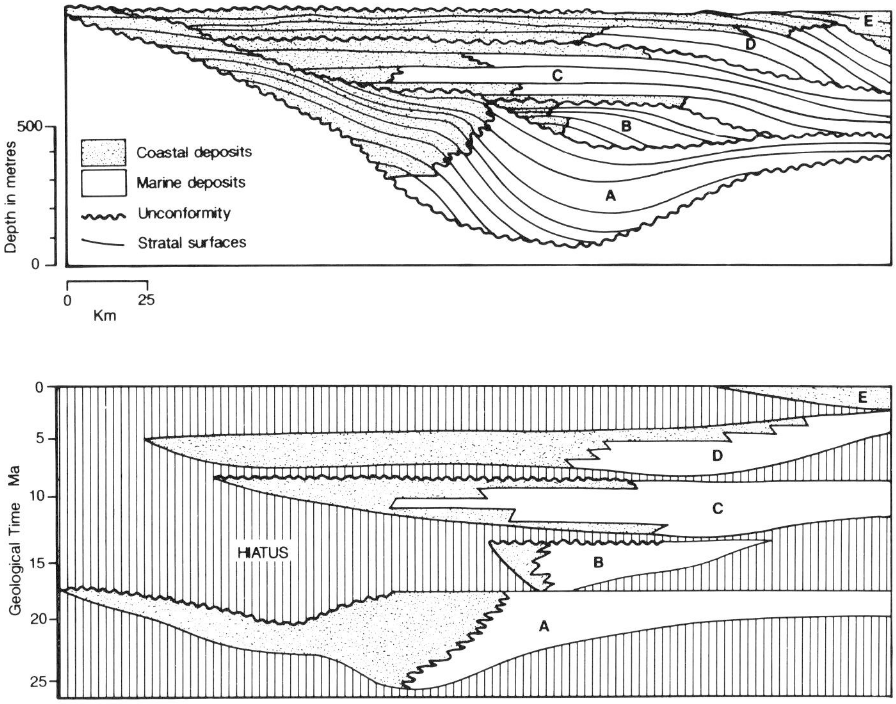

Fig. 4. Diagrammatic representation of a stratigraphical cross section and its chronostratigraphical chart. The upper drawing shows a stratigraphical cross section derived from seismic reflection data. Sequences A, B, C, D and E are separated by unconformities which form sequence boundaries. Stratal surfaces/reflectors are traceable in different lithologies within each sequence. Age control for the chronostratigraphical chart (lower drawing) is provided by available well or surface sections and the interrelationships between stratal surfaces/reflectors. Based on Vail *et al* 1977.

separated from each other by continent-wide unconformities. Subsequent development in the understanding of basin subsidence, particularly its episodicity and its punctuation by (plate) tectonic and possibly sea-level events, together with the advent of seismic stratigraphy, have reinforced the view that the sequence concept is valuable in stratigraphical analysis (Wilgus *et al.* 1988).

Sequence analysis is essentially interpretative and thus incorporates a significant element of subjectivity. It is therefore less amenable to strict standards of typology, nomenclature and classification than is lithostratigraphy, which is primarily descriptive. Guidelines are therefore based on usage rather than on attempted formalizations such as those of the International Subcommission on Stratigraphic Classification (Salvador 1987*a*), though the reader should be aware of these. The use of the word 'sequence' might be regarded by some as unfortunate, but 'succession' is still available for more general use.

A (depositional) **sequence** is a package of strata bounded above and below by unconformities of regional (e.g. basin-wide) extent (or their correlative conformities where the succession is conformable). As such they can also be thought of in terms of process; thus a depositional sequence is the stratigraphical record that accumulated between significant changes of tectonic (or perhaps eustatic) regime, within an overall episode of basin subsidence.

A full hierarchy of sequence stratigraphical terms is neither necessary nor desirable at present. Nevertheless, one larger and smaller scale unit can be usefully recommended, as being of sufficiently different interpretative significance, as well as scale, from the sequence itself.

A **megasequence** is a grouping of sequences that were deposited between major basin-forming events; it is equivalent to the concept of a basin as a genetic unit (whereby basins may be stacked one on another) rather than simply as a geographical unit.

A **sub-sequence** is a unit of lower rank, bounded by unconformities that are not basin-wide and thus have local rather than regional significance. For example, the transgressive basal and progradational top parts of a single sequence may locally be separated by a disconformity which can be regarded as dividing it into two sub-sequences.

At present the sequence is the primary unit because it can be directly recognized on seismic reflection profiles.

Cycles in stratigraphy

The effects of transgressive–regressive cycles in the geological record have influenced much stratigraphical work and thought. Important stratigraphical boundaries, particularly those between the systems, have been explicitly based on major transgressive–regressive cycles, initially founded on local observations but later extended to infer transgressive–regressive events at a global scale. Stratigraphy has gone through a phase during which it was unacceptable to admit the global extent of such events, but work in several essentially independent disciplines has combined to bring the topic of cyclicity and its possible global importance to prominence once more. Noteworthy amongst these is seismically based sequence stratigraphy, discussed above. The interpretation put on the inferred correlatability of sequences at a wide range of scales is that transgressive–regressive cycles might have a major controlling influence on stratigraphy world-wide at a wide range of frequencies.

On the longest time-scales, transgressive–regressive cycles are though to be caused by the repeated plate tectonic separation and reassembly of Pangaea. On shorter time-scales, transgressive–regressive cycles are associated with glacio-eustasy in the later Neogene and probably the Permo-Carboniferous. Controversy continues to surround the relative importance of tectonic effects at high frequencies (say, less than 5 million years), and of high frequency sea-level change, and whether or not it could be glacio-eustatically driven, at other apparently non-glacial times in Earth history.

The probability that astronomical influences (Milankovitch theory) have a causal effect on the episodic major glaciations of the Pleistocene has increased with the demonstration that the frequency characteristics of oxygen isotope ratio variations (from piston-cored deep ocean sediments) perfectly match the variations in received solar energy predicted from computation of the earth's orbital complexities (Hays *et al.* 1976). However, despite the apparent causal connection, the transfer mechanism from astronomical input to stratigraphical signal is not yet understood. Nevertheless, well-documented claims of orbitally-forced rhythmicity in pre-Pleistocene strata (e.g. Berger *et al.* 1984; House 1985; Weedon 1985) suggest that a high frequency signal may be more pervasive of the total stratigraphical record than has been considered likely hitherto. Indeed, the various claims of very long distance correlation of thin beds or groups of beds may now have a good theoretical basis, giving high resolution stratigraphical correlation a newly challenging and optimistic outlook.

Developments related to cycles in stratigraphy sometimes give rise to new terminology. For example, Ramsbottom (1977) favoured the terms synthem and mesothem for transgressive–regressive cycles in the British Carboniferous succession. Although these terms have not been widely used, the International Subcommission on Stratigraphic Classification (Salvador 1987*a*) has affirmed their use in the context of unconformity-bounded stratigraphical units. It may be premature to recommend use of these terms at this time.

Event stratigraphy

Events may be defined broadly as comparatively rare and geologically short-lived occurrences which have left some trace in the rock records, and which may be employed for correlation. That is, correlation is achieved by means of inferred geological events rather than any intrinsic characteristics of the rocks themselves (Ager 1973). Events thus include happenings such as volcanic eruptions, earthquakes, floods and storms, mass-movement and turbidity flows, climatic events and sea-level changes. They range in scale from global catastrophes, such as the inferred meteorite impacts which have been associated with iridium-rich levels and major extinctions (Silver & Schutz 1982), to basin-wide or relatively local mortality events or 'blooms' of certain fossil taxa. One of the best known examples of the application of events to correlation is the use of ash-fall horizons for correlation between Upper Carboniferous basins in lieu of biostratigraphical evidence (Bouroz 1967). Also in the Carboniferous, Ramsbottom (1977, 1981) has recognized a series of eustatic events and rhythms, and has defined a parallel set of chronostratigraphical stages. His mesothems (see above) were units of stage rank bounded by unconformities in cratonic areas, but with their limits defined in conformable successions (Holland *et al.* 1978, Fig. 3). The theory and application of these event correlations have recently been discussed (Walkden 1987).

Certain transgressive–regressive 'events' have been taken to be widespread and synchronous (e.g. see Hallam 1978), the most spectacular example being a transgression used to define the Jurassic–Cretaceous boundary, in the Boreal realm at least (Casey 1973; Rawson *et al.* 1978). Ager (1973) has urged caution in the use of event correlation in the absence of evidence for time correlations provided by evolving fossil lineages. The limitations of studies of the rock record and the identification of synchronous events divorced from any biostratigraphical control are obvious (Talbot 1973; Cope *et al.* 1980*a*): there is a need to relate precisely located sedimentary events to a rigorous biostratigraphical framework (Townson 1975; Wimbledon 1987).

Microfaunal and microfloral fossil groups offer some of the best possibilities for precise long-range correlations, giving evidence of oceanic anoxic, eustatic or high productivity events. Perhaps the best prospects for high resolution event stratigraphy come from geochemical techniques, for instance by the use of stable isotopes, such as oxygen, carbon, sulphur and strontium (Odin *et al.* 1982; Shackleton 1985). The use of the first of these in deep sea cores has revolutionized correlation in the Pleistocene (Emiliani 1955); a relatively complete climatic record, based on $^{18}O/^{16}O$ ratios, being used as the standard with which fragmentary continental sequences are compared (Shackleton & Opdyke 1973).

Basement complexes and terranes

In areas dominated by igneous or high-grade metamorphic rocks it may not be possible to erect a meaningful lithostratigraphy (Salvador 1987*b*). Instead, such an area is often described as a complex, a typical example being provided in NW Scotland by the Lewisian Complex, the different tectonic units within which (e.g. Scourian and Laxfordian gneisses) are distinguished on the basis of distinctive isotopic, metamorphic and structural characteristics. No word is used formally to describe subdivisions recognizable within a complex, although the term 'unit' can be useful. Similarly no word is in general usage to describe a number of complexes, although the term supercomplex could be appropriate here.

The concept of **suspect terranes** has emphasized how areas with unrelated stratigraphical sequences are commonly juxtaposed in orogenic belts by movement along transcurrent faults (e.g. Soper *et al.* 1989). The modern use of the term suspect terrane refers to an area bounded by faults and possessing a distinctive stratigraphy that is markedly different from those of neighbouring terranes. The tectonic boundaries may exhibit brittle and/or ductile (shear zone) fault rocks. If it can be proved that a terrane has been displaced large distances and is not indigenous to the cratonic margin, then such an area is often called an **exotic terrane**. Subsequent to the 'docking' of two such terranes, both should exhibit some stratigraphical connection such as a sedimentary overstep sequence or an igneous intrusion cutting the boundary (a 'stitching pluton').

Some areas of complicated and poorly understood geology may, with further work, be shown to possess a recognizable lithostratigraphy. An example is provided by the Moine schists NW of the Great Glen fault, a metamorphic complex previously given several informal terms such as assemblage or succession, but now referred to as the Moine Supergroup. It is recommended that the use of informal terms (assemblage, succession, division) in such areas should be avoided in favour of complex. In contrast to the Moine example, further study of other complexes can reveal the existence of unrelated terranes. An example of this is provided by the 'Mona complex' of Anglesey, which contains tectonometamorphic units (e.g. blueschists) isolated by mylonite zones and possessing no proven stratigraphical connection with adjacent areas. No formal stratigraphical term exists to describe such areas containing several highly disparate terranes, although informal use of 'collage' is frequently made. Use of the term superterrane (analogous to supergroup) is increasingly common (e.g. Gibbons 1990). Once a terrane collage has been identified, further work should focus upon characterizing the stratigraphical signature and tectonic setting of each terrane, determining the amount and sense of relative displacement, and identifying the time of docking.

Some problems of stratigraphical usage

Figure 1 illustrates some of the hierarchical relationships between stratigraphical divisions that have been applied in the Penarth area of South Wales (the type area of the Penarth Group). The figure can also be used to demonstrate some of the difficulties involved in, and potential consequential effects of, stratigraphical classification and nomenclature. Details of the sections are taken from Warrington *et al.* (1980) and Waters & Lawrence (1987).

Problems of lithostratigraphical classification in this region stem from the fact that the sediments making up the sequence show a gradual transition from a dominantly non-marine environment (Mercia Mudstone Group) to a fully marine one (Blue Lias). Rapid alternations of thin beds, and repetitions of 'packets' of sediments formed in similar environments, make the definition of formation boundaries difficult even in continuous local quarry or cliff exposures. The accurate geological mapping of these formation boundaries away from local well-exposed outcrops in areas with a full, thick alternating sequence, is even more difficult.

Although the Blue Anchor Formation (Mercia Mudstone Group) and the Westbury Formation (Penarth Group) meet the criterion of internal lithological consistency reasonably well, the Lilstock Formation does so less well. The latter is divided into an underlying Cotham Member (calcareous siltstones, mudstones, limestones) and an overlying Langport Member (limestones, mudstones) each of which, arguably, could be regarded as a formation on the grounds of lithological consistency. Such a proposal, however, would pose problems either of redundancy, or of hierarchical placing of the term Lilstock Formation (Fig. 1). This could be seen as an example of a stratigraphical sequence being forced into a classification strait-jacket.

With regard to the criterion of mappability, this depends, of course, on factors such as scale of mapping and the continuity of exposure. At suitable scale in quarry and cliff exposures it is possible to trace and map locally all the named formations and even the members. Broader scale, and regional mapping present problems particularly in poorly exposed terrain such as that typical of much of Britain. For example, the Blue Anchor Formation presents accurate boundary tracing problems because at the base are alternating green mudstone and red siltstone beds, and at the top, alternating green mudstone and dark grey mudstone beds. The Westbury and Lilstock formations, although recognizable as formations where well exposed, are, in similar circumstances of broad-scale mapping in poorly exposed terrain, of necessity lumped together and mapped as the Penarth Group. Thus, the two principal criteria applying to lithostratigraphical units (internal lithological consistency and mappability) can present additional but little-recognized problems of reconciliation between local, detailed small-scale work in well-exposed terrain (necessary in order to define formations) and the realities and requirements of regional, broad-scale mapping work in poorly-exposed terrain.

Such difficulties are further compounded in the British situation by historical usage. The term Blue Lias (Fig. 1), for example, in constant use since William Smith's first application in 1799 (Cox 1942), was formally defined as the Blue Lias Formation (type section at Saltford railway cutting) by Cope *et al.* (1980*a*). The formal definition (Waters & Lawrence 1987) of the Porthkerry Formation and the St Mary's Well Bay Formation (Fig. 1) within the Blue Lias (by implication in this usage, of group status) would appear to confuse the hierarchy. However, although the Blue Lias does meet the criteria of internal lithological consistency and mappability, so do the discrete, recognizable, internal divisions within it. Appropriate local usage has clashed with appropriate regional usage.

Unless the undesirable proliferation of new hierarchical rankings such as 'superformation' or 'subgroup' were to be allowed, the consequential effects from local, detailed work and definition could entail the loss of regionally useful terms of higher rank. For example, the Blue Lias itself (whether of formation or group status), is part of the Lower Lias, the lowest of the three major divisions of the Lias. This tripartite division of the Lias into Lower Lias, Middle Lias and Upper Lias, applied since the time of Phillips (1829) and still of great practical utility on the regional and super-regional scale, is in danger of being lost by use of the term 'Lias Group', commonly defined by the grouping of locally recognized formations. So that practical utility on the regional or super-regional scale is not subordinated to rigid, local application of 'rules', it would be of the greatest practical utility in this particular case to reinstate use of the terms Lias, Lower Lias, Middle Lias and Upper Lias without addition of a hierarchical or rank term.

The ammonite biozones and sub-biozones shown in Fig. 1 are well established in the area concerned; they are essentially based upon successive first appearances of ammonites and are thus a type of range or interval biozone. Although they are shown in the diagram as biostratigraphical units, the fact that similar sequences of ammonite faunas are found over wide areas, often intercontinental, has caused many workers to regard them as chronostratigraphical zones and sub-chronostratigraphical zones, a usage that blurs the distinction between biostratigraphical and chronostratigraphical divisions. Until properly proposed and delimited using chronostratigraphical criteria (i.e. with chronostratigraphical stratotype section and point) they should not be regarded as chronozones.

Usage of formal and informal stratigraphical names

Confusion about the conventions involving use of initial capital letters with stratigraphical names is easily overcome by the following simple rule. All words in every formally named stratigraphical unit begin with capital letters except where specific names of fossils are used. For example, the

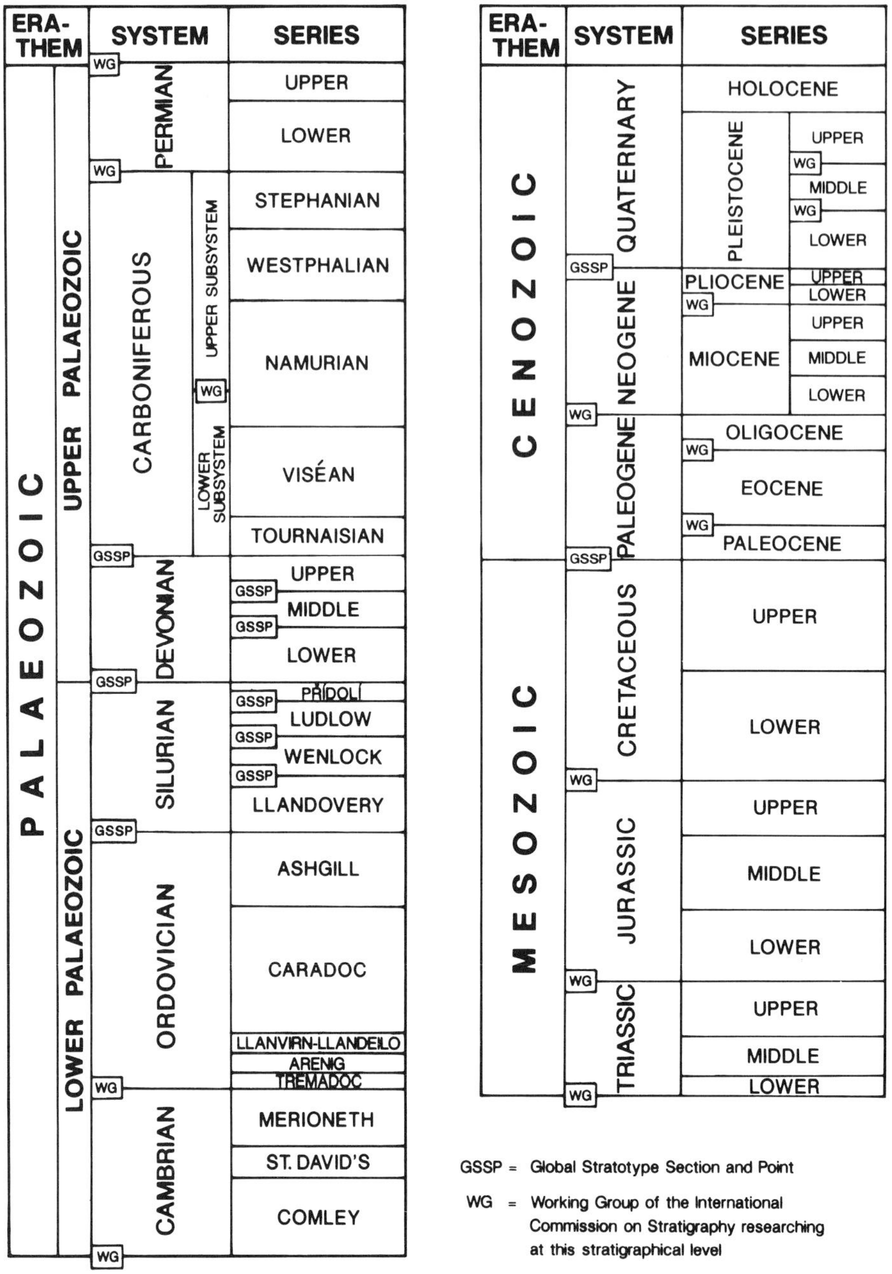

Fig. 5. Phanerozoic chronostratigraphical divisions of Erathem, System and Series rank currently used in the British Isles.

Alsatites liasicus Biozone is referred to the Hettangian Stage of the Jurassic System.

Some Phanerozoic Period/System and Epoch/Series names and some boundaries have been formally proposed and agreed upon by the International Commission on Stratigraphy of IUGS (Cowie & Bassett 1989) (see Fig. 5). Tables of generally accepted but informal Age/Stage names are given in Harland *et al.* (1990), Odin (1982*a*, *b*, 1984), Palmer (1983) and Snelling (1985); these references include isotopic dates of Age/Stage boundaries but do not agree on all Age/Stage names or numerical ages of boundaries.

Capitalized words referring to time (Early, Mid-, Late) must be used in a precise way about formally defined divisions, whereas uncapitalized words (early, mid-, late) can be used less precisely. Thus, 'early Jurassic' may be used less precisely than 'Early Jurassic' which refers specifically to the Early Jurassic Epoch, that is, the time period from the Hettangian Age to the Toarcian Age inclusive.

Words referring to relative vertical position (lower, middle, upper) in a chronostratigraphical division can be used either formally or informally. For example, Lower Cambrian refers to the Lower Cambrian Series, whereas lower Cambrian may mean in a general and imprecise sense the lower part of the Cambrian. Usage of lower, middle and upper in lithostratigraphical or biostratigraphical units is essentially informal (e.g. lower Coalbrookdale Formation; lower *lundgreni* Biozone), although some British lithostratigraphical terms such as Upper Old Red Sandstone and Lower Lias are given thus because of long-established, historical practice.

It is helpful to remember that many stratigraphical terms which are now commonly used as nouns (Silurian; Pliensbachian) were originally adjectives (Silurian System; Pliensbachian Stage). Treatment of the names as adjectives by users would dispel much confusion, particularly in the case of verbal presentation when it is impossible to make the distinction between capitalized and uncapitalized words.

A. W. A. Rushton and A. Whittaker publish with the permission of the Director, British Geological Survey (NERC). The authors are grateful to M. G. Bassett and Staff at the National Museum of Wales for much advice and practical help with the text figures.

References

AGER, D. V. 1973. *The nature of the stratigraphical record.* Macmillan, London.

ANON. 1988. Personal profile, Robert Ginsburg-global sedimentologist. *Episodes,* **11,** 57–58.

BERGER, A. L., IMBRIE, J., HAYS, G. *et al.* (eds). 1984. *Milankovitch and Climate: Understanding the Response to Astronomical Forcing.* Reidel, Dordrecht.

BOUROZ, A. 1967. Correlation des tonsteins d'origin volcanique entre les bassins houilliers de Sarre-Lorraine et du Nord-Pas de Calais. *Compte Rendu Hebdomadaire des Seances de l'Academie des Sciences. Serie D Science Naturelle, Paris,* **264,** 2729–2732.

CASEY, R. 1973. The ammonite succession at the Jurassic-Cretaceous boundary in eastern England. *In:* CASEY, R. & RAWSON, P. F. (eds) *The Boreal Lower Cretaceous.* Geological Journal Special Issue No. 5. Seel House Press, Liverpool.

COPE, J. C. W., GETTY, T. A., HOWARTH, M. K. *et al.* 1980*a*. *A correlation of Jurassic rocks in the British Isles. Part One: Introduction and Lower Jurassic.* Geological Society of London Special Report No. **14**.

——, DUFF, K. L., PARSONS, C. F. *et al.* 1980*b*. *A correlation of Jurassic rocks in the British Isles. Part Two: Middle and Upper Jurassic.* Geological Society of London Special Report No. **15**.

COWIE, J. W. & BASSETT, M. G. 1989. Global Stratigraphic Chart. International Union of Geological Sciences. *Episodes,* **12** (supplement).

COX, L. R. 1942. New light on William Smith and his work. *Proceedings of the Yorkshire Geological Society,* **25,** 1–99.

DEEGAN, C. E. & SCULL, B. J. 1977. *A standard lithostratigraphic nomenclature for the Central and Northern North Sea.* Report of the Institute of Geological Sciences, No. **77/25**.

DORE, A. G., VOLLSET, J. & HAMAR, G. P. 1985. Correlation of the offshore sequences referred to the Kimmeridge Clay Formation–relevance to the Norwegian sector. *In:* THOMAS, B. W. *et al.* (eds) *Petroleum Geochemistry in Exploration of the Norwegian Shelf.* Graham and Trotman, London.

EMILIANI, C. 1955. Pleistocene temperatures. *Journal of Geology,* **63,** 538–578.

GIBBONS, W. 1990. Transcurrent ductile shear zones and the dispersal of the Avalon superterrane. *In:* D'LEMOS, R. S., STRACHAN, R. A. & TOPLEY, C. G. (eds) *The Cadomian Orogeny.* Geological Society, London, Special Publication, **51,** 407–423.

GREUTER, W., BURDET, H. M., CHALONER, W. G. *et al.* 1988. *International Code of Botanical Nomenclature.* Koeltz Scientific Books, Koenigstein, Germany.

HAILWOOD, E. A. 1989. *Magnetostratigraphy.* Geological Society of London Special Report No. **19**.

HALLAM, A. 1978. Eustatic cycles in the Jurassic. *Palaeogeography, Palaeoclimatology, Palaeoecology,* **35,** 1–44.

HANCOCK, J. M. 1977. The historic development of concepts of biostratigraphic correlation. *In:* KAUFFMAN, E. & HAZEL, J. E. (eds) *Concepts and methods of biostratigraphy.* Dowden, Hutchinson and Ross Inc., Stroudsburg, 3–22.

HARLAND, W. B., ARMSTRONG, R. L., COX, A. V., *et al.* 1990. A geologic time scale 1989. Cambridge University Press.

HAYS, J. D., IMBRIE, J. & SHACKLETON, N. J. 1976. Variations in the Earth's orbit: pacemaker of the ice ages. *Science,* **194,** 1121–1132.

HOLLAND, C. H. 1986. Does the golden spike still glitter? *Journal of the Geological Society, London,* **143,** 3–21.

——, AUDLEY-CHARLES, M. G., BASSETT, M. G. *et al.* 1978. *A guide to stratigraphical procedure.* Geological Society of London Special Report No. **10**.

—— & BASSETT, M. G. 1988. Discussion on biostratigraphical correlation and the stages of the Llandovery. *Journal of the Geological Society, London,* **145,** 881–882.

HOUSE, M. R. 1985. A new approach to an absolute time-scale from measurements of orbital cycles and sedimentary microrhythms. *Nature,* **315,** 721–725.

ODIN, G. S. (ed.) 1982*a*. *Numerical dating in stratigraphy.* Wiley, New York. 2 volumes.

—— 1982*b*. The Phanerozoic time-scale revisited. *Episodes,* **5,** 3–9.

—— 1984. The numerical age of system, series, and stage boundaries of the Phanerozoic column. *In:* SEIBOLD, E. & MENLENKAMP, J. D. (eds) *Stratigraphy Quo Vadis?* American Association of Petroleum Geologists, Studies in Geology, **16,** 61–64.

——, RENARD, M. & GRAZZINI, C. V. 1982. Geochemical events as a means of correlation. *In:* ODIN, G. S. (ed.) *Numerical dating in stratigraphy.* Wiley, New York, 37–71.

PALMER, A. R. 1983. The Decade of North American Geology 1983 time scale. *Geology,* **11,** 503–504.

PAYTON, C. E. (ed.) 1977. *Seismic stratigraphy—applications to hydrocarbon exploration.* American Association of Petroleum Geologists, Memoir, **26**.

PHILLIPS, J. 1829. *Illustrations of the Geology of Yorkshire; or a description of the strata and organic remains of the Yorkshire coast: accompanied by a geological map, sections, and plates of the fossil plants and animals.* J. Phillips, York. Printed for the author by Thomas Wilson.

RAMSBOTTOM, W. H. C. 1977. Major cycles of transgression and regression (mesothems) in the Namurian. *Proceedings of the Yorkshire Geological Society,* **41,** 261–291.

—— (ed.) 1981. *Field guide to the boundary stratotypes of the Carboniferous stages in Britain.* IUGS Subcommission on Carboniferous Stratigraphy, Leeds.

RAWSON, P. F., CURRY, D., DILLEY, F. C., *et al.* 1978. *A correlation of Cretaceous rocks in the British Isles.* Geological Society of London, Special Report No. **9**.

RHYS, G. H. (compiler) 1974. *A proposed standard lithostratigraphic nomenclature for the southern North Sea and an outline structural nomenclature for the whole of the (UK) North Sea.* Report of the Institute of Geological Sciences No. **74/8**.

—— 1975. *A proposed standard lithostratigraphic nomenclature for the southern North Sea. In:* WOODLAND, A. W. (ed.) *Petroleum and the Continental Shelf of North-West Europe, Volume 1 Geology.* Applied Science Publishers Ltd, London, 151–162.

RIDE, W. D. L., SABROSKY, C. W., BERNARDI, G. & MELVILLE, R. V. 1985. *International Code of Zoological Nomenclature.* International Trust for Zoological Nomenclature, British Museum (Natural History), London. University of California Press.

SALVADOR, A. (Chairman) 1987*a*. Unconformity-bounded stratigraphic units. International Subcommission on Stratigraphic Classification. *Geological Society of America Bulletin,* **98,** 232–237.

——, 1987*b*. Stratigraphic classification and nomenclature of igneous and metamorphic rock bodies. *Geological Society of America Bulletin,* **99,** 440–442.

SHACKLETON, N. J. 1985. Neogene oxygen and carbon isotope stratigraphy. *In:* SNELLING, N. J. (ed.) *The Chronology of the Geological Record.* Geological Society, London, Memoir, **10,** 206–208.

—— & OPDYKE, N. D. 1973. Oxygen isotope and palaeomagnetic stratigraphy of Equatorial Pacific core V28-238. *Quaternary Research,* **3,** 39–55.

SILVER, L. T. & SCHUTZ, P. H. (eds) 1982. *Geological implications of impacts of large asteroids and comets on the earth.* Geological Society of America Special Paper 190.

SLOSS, L. L. 1963. Sequences in the cratonic interior of North America. *Geological Society of America Bulletin,* **74,** 93–114.

SNELLING, N. J. (ed.) 1985. *The Chronology of the Geological Record.* Geological Society of London, Memoir, **10**.

SOPER, N. J., GIBBONS, W. & MCKERROW, W. S. 1989. Displaced terranes in Britain and Ireland. *Journal of the Geological Society, London,* **146,** 365–367.

TALBOT, M. R. 1973. Major sedimentary cycles in the Corallian beds (Oxfordian) of southern England. *Palaeogeography, Palaeoclimatology, Palaeoecology,* **14,** 293–317.

TEMPLE, J. T. 1988. Biostratigraphical correlation and the stages of the Llandovery. *Journal of the Geological Society, London,* **145,** 875–879.

TOWNSON, W. G. 1975. Lithostratigraphy and deposition of the type Portlandian. *Journal of the Geological Society, London,* **131,** 619–638.

VAIL, P. R. 1987. Seismic stratigraphy interpretation procedure. *In:* BALLY, A. W. (ed.) *Atlas of Seismic Stratigraphy.* American Association of Petroleum Geologists, Studies in Geology, **27,** 11–14.

——, MITCHUM, R. M. & THOMPSON, S. 1977. Seismic Stratigraphy and Global Changes of Sea Level, Part 3: Relative Changes of Sea Level from Coastal Onlap. *In:* PAYTON, C. E. (ed.) *Seismic stratigraphy—applications to hydrocarbon exploration.* American Association of Petroleum Geologists, Memoir, **26,** 63–81.

VAN WAGONER, J. C., MITCHUM, R. M., POSAMENTIER, H. W. & VAIL, P. R.

1987. Key definitions of sequence stratigraphy. *In*: BALLY, A. W. (ed.) *Atlas of Seismic Stratigraphy*. American Association of Petroleum Geologists, Studies in Geology, **27,** 11–14.

VAUGHAN, A. 1905. The palaeontological succession in the Carboniferous Limestone of the Bristol area. *Quarterly Journal of the Geological Society of London,* **61,** 181–307.

WALKDEN, G. M. 1987. Sedimentary and diagenetic styles in late Dinantian carbonates. *In*: MILLER, J., ADAM, A. E. & WRIGHT, V. P. (eds) *European Dinantian Environments*. Wiley & Sons, Chichester.

WARRINGTON, G., AUDLEY-CHARLES, M. G., ELLIOTT, R. E. *et al.* 1980. *A correlation of Triassic rocks in the British Isles*. Geological Society of London Special Report No. **13**.

WATERS, R. A. & LAWRENCE, D. J. D. 1987. *Geology of the South Wales Coalfield, Part III, the country around Cardiff*. Memoir of the British Geological Survey.

WEEDON, G. P. 1985. Hemipelagic shelf sedimentation and climatic cycles: the basal Jurassic (Blue Lias) of South Britain. *Earth and Planetary Science Letters,* **76,** 321–335.

WHITTAKER, A., HOLLIDAY, D. W. & PENN, I. E. 1985. *Geophysical logs in British stratigraphy*. Geological Society of London Special Report No. **18**.

WILGUS, C. K., HASTINGS, B. S., KENDALL, C. G. S. C. *et al.* (eds) 1988. *Sea-level changes: an integrated approach*. Society of Economic Paleontologists and Mineralogists, Special Publication, **42**.

WIMBLEDON, W. A. 1987. Rhythmic sedimentation in the late Jurassic–early Cretaceous. *Proceedings of the Dorset Natural History and Archaeological Society,* **135,** 127–133.

WOOD, C. J. & SMITH, E. G. 1978. Lithostratigraphical classification of the Chalk in north Yorkshire, Humberside and Lincolnshire. *Proceedings of the Yorkshire Geological Society,* **42,** 263–287.

WOOLLAM, R. & RIDING, J. B. 1983. *Dinoflagellate cyst zonation of the English Jurassic*. Report of the Institute of Geological Sciences, No. **83/2**.

Received 27 November 1990; revised typescript accepted 20 March 1991